GRAND CANYON
WILD
RIM TO RIVER
LIFE

Grand
Canyon
Association

MIKE BUCHHEIT

Grand Canyon Association
PO Box 399
Grand Canyon, Arizona 86023
(800) 858-2808
www.grandcanyon.org

Printed in the United States of America
Edited by Pam Frazier
Designed by Amanda Summers Design
graphy of Grand Canyon habitats by Tom Bean

ISBN 978-1-934656-41-9-48-8

Library of Congress Cataloging-in-Publication Data

Lamb, Susan, 1951-
Grand Canyon wildlife : rim to river / by Susan Lamb.
pages cm
Includes index.
ISBN 978-1-934656-48-8
1. Canyon animals--Arizona--Grand Canyon National Park. 2. Habitat (Ecology)--Arizona--Grand Canyon National Park. 3. Life zones--Arizona--Grand Canyon National Park. 4. Natural history--Arizona--Grand Canyon National Park. 5. Merriam, C. Hart (Clinton Hart), 1855-1942. 6. Grand Canyon National Park (Ariz.)--Environmental conditions. 7. Grand Canyon National Park (Ariz.)--Description and travel. I. Title.
QL162.L25 2014
577.09791'32--dc23

2013018260

The Grand Canyon Association is the National Park Service's official nonprofit partner, raising private funds to benefit Grand Canyon National Park, operating retail stores and visitor centers within the park, and providing premier educational opportunities about the natural and cultural history of Grand Canyon. Proceeds from the sale of this publication will be used to support research and education at Grand Canyon National Park.

DEDICATION

In memory of Grand Canyon Wildlife Biologist Eric York 1970–2007, an accomplished and ardent advocate for wildlife.

COVER: CONDOR, TOM JONES; SHEEP, NATIONAL PARK SERVICE; RINGTAILS, MINDEN PICTURES/SUPERSTOCK
TITLE PAGE: TOM BEAN COURTESY OF LARRY STEVENS
BACK COVER: BACKGROUND,NATIONAL PARK SERVICE; BUTTERFLY, TOM BEAN COURTESY OF LARRY STEVENS

CONTENTS

In the end we will conserve only what we love. We will love only what we understand. We will understand only what we are taught."

—Dr. Baba Dioum

JOHN PITCHER/ THINKSTOCK

INTRODUCTION

The Different Worlds of Grand Canyon Wildlife

We are more aware of wild creatures when we sleep out under the night sky. Of course we see animals during the day—a mule deer browsing in a clearing, a rock squirrel scampering along the rim of the Grand Canyon or, if we're lucky, a condor or a bighorn sheep. But when we close our eyes to wait for sleep, we hear them.

Wild sounds aren't meant for us but they tell us something important. Crickets chirping on a warm summer night remind us of the countless creatures we rarely see. Canyon tree frogs chorusing over the roar of the river, a great horned owl hooting from a tree, the scrambled yowls of coyotes; all prompt us to remember that this is their world too.

Theirs isn't exactly the same world as ours. Many animals perceive what is around them in ways that are mysterious to us. Migrating birds sense the earth's magnetic field. Bats calculate distance and direction with sonar. Butterflies taste with their feet.

Often, animal senses are tuned to a different frequency than ours: bees see ultraviolet, rattlesnakes detect infrared radiating from their prey, coyotes hear high-pitched sounds made by mice brushing against grass.

How many hearts with warm red blood in them are beating under cover of the woods, and how many teeth and eyes are shining! A multitude of animal people, intimately related to us, but of whose lives we know almost nothing, are as busy about their own affairs as we are about ours.

—John Muir,
Our National Parks, 1909

JOHN MUIR (1838–1914)

When we try to pick out anything by itself we find that it is bound fast by a thousand invisible cords that cannot be broken, to everything in the universe.

—from the Journal of John Muir, July 27, 1869

John Muir wrote knowledgeably of the scientific reasons for conserving Grand Canyon. Yet his charming portraits of its wildlife "its multitude of lesser animals, well-dressed, clear-eyed, happy little beasts" and poetic descriptions of their environment recruited more popular support than did his sober arguments.

Influential people were affected by Muir's heartfelt love of nature too. In 1896, Muir took a walk along the South Rim with Gifford Pinchot, future director of the U.S. Forest Service. Pinchot noted with astonishment that "when we came across a tarantula, he wouldn't let me kill it. He said it had as much right there as we did."

Muir also went camping with Theodore Roosevelt shortly after the president had visited Grand Canyon. The two men talked long into the night about conserving wild habitat for its own sake. With the encouragement of the American people, Roosevelt later defied a slow-moving Congress to create the Grand Canyon Game Reserve and in 1908, used the Antiquities Act to declare Grand Canyon a national monument.

Although Muir didn't live to see it, his eloquent appeals enlisted wide public support for the legislation that established Grand Canyon National Park in 1919.

Many animals, like bobcats, thrive in a range of habitats, while others, such as tassel-eared squirrels, are so interwoven with a certain environment that they are seldom found anywhere else. Some animals are wedded to their own time and place whether it is a desert or forest, riverbank or cliff. A single summer in a flowery meadow can be a whole lifetime for a beetle, while a desert tortoise may nibble scrub for a century.

Raccoons forage at night and cottontails browse at dusk and dawn. Ground squirrels hibernate in winter; spadefoot toads estivate in summer. We mostly keep our feet on the ground, but other animals soar in air, swim in streams, tunnel underground, or hang upside down in caves.

It is said that 70 percent of what humans observe is what we see, 20 percent is what we hear, and only a little comes to us through our other senses of taste, smell, and touch. Yet the sun's warmth or the scent of damp earth can affect us in hidden ways. Neurologists know that all our senses influence our paleo-mammalian brain, which is also called our seat of emotion, the bridge to our memories, and the clock that sets our circadian rhythms. We are more like other animals than we realize.

Bobcat — TOM BEAN

Opposite: Many Grand Canyon creatures are active mostly at night.
TYLER NORDGREN

THE MOUNTAIN LION—A Keystone Species

A thing is right when it tends to preserve the integrity, stability, and beauty of the biotic community. It is wrong when it tends otherwise.

—Aldo Leopold, *Journal of Forestry*, 1933

It is close to midnight on April 6, 2005. There is no moon, but countless stars shimmering in the cold black sky lend a silvery gleam to the sunken landscape of Grand Canyon.

P05 pauses below the South Rim near Hermits Rest. Like those of all cats, her enormous amber eyes take in a wide-angle view and magnify the starlight with mirror-like *tapeta lucida*—"shining tapestries"—behind their retinas.

Biologists caught and collared P05 at Dripping Springs, then tracked her nosing along the South Rim for just a few weeks. They know she is old enough to mate but that there are at least two older female mountain lions in the area.

P05 begins to pad downslope, her huge paws soundless on the ground, her long and heavy tail keeping her balanced as it does when she leaps dozens of feet toward her prey. At 3:00 a.m. she is at Lookout Point. She keeps going, her power and agility evident in the muscles packed under her loose hide. She soon reaches the Colorado River and swims across, reaching 94-mile Creek about 5:00 a.m. as a sliver of moon finally rises above the horizon. Dawn lightens the sky as she makes her way up to the North Rim.

Two months later, her collar stops sending signals and the park biologist loses track of her. It's frustrating; mountain lions are difficult to capture,

FUSE/THINKSTOCK

and the tracking collar that pinpointed her daily movements was expensive. Still, P05 has demonstrated that mountain lions do cross the Colorado River, which is important for healthy genetic exchange.

Puma concolor—the mountain lion—is a keystone species for the Grand Canyon area. Like the single keystone in a masonry arch, a mountain lion helps stabilize the much greater numbers of deer and other plant-eaters. Without keystone predators to keep them in check, irruptions of herbivores can devastate the grasses, shrubs, and trees on which they depend not only for food, but also for shelter.

Of the more than ninety mammal species at the Grand Canyon, mountain lions are among only a few that may be found from rim to river, north to south, east to west. Animals more typically have a specific habitat based on their needs for water, food, and shelter, and on their tolerance for heat or cold.

ALDO LEOPOLD (1887–1948)

The last word in ignorance is the man who says of an animal or plant: "What good is it?"
—from the notes of Aldo Leopold

In 1909, Aldo Leopold came to Arizona to work for the U.S. Forest Service. His duties included shooting predators to ensure a "shootable surplus" of game and to protect livestock.

One fateful day, Leopold shot a mother wolf. Watching "a fierce green fire dying in her eyes," he suddenly felt that even wolves must have a purpose in nature. When he saw that an increasing number of deer were devastating the vegetation, he began to question the policy of eliminating predators.

Fifteen years later, Grand Canyon Game Reserve rangers realized that removing livestock and shooting mountain lions, bobcats, and coyotes had led not only to more deer but also to badly degraded vegetation on the North Rim. They opened the reserve to hunting and supported an attempt to herd deer across the canyon to the South Rim. The "Kaibab Deer Drive" turned into a fiasco when the deer simply jumped out of the way.

COURTESY OF THE ALDO LEOPOLD FOUNDATION, WWW.ALDOLEOPOLD.ORG

Leopold used the Grand Canyon episode to point out how reducing predators to benefit herbivores can severely damage habitat. He proposed a land ethic embracing whole ecosystems: soil, water, and all plants and animals.

Although ecologists now recognize other reasons for swings in wildlife populations, Leopold's statement of the vital role of predators remains an important ecological principle.

C. HART MERRIAM AND THE LIFE ZONE CONCEPT

In descending from the plateau level to the bottom of the cañon a succession of temperature zones is encountered equivalent to those stretching from the coniferous forests of northern Canada to the cactus plains of Mexico. They result from the combined effects of altitude and slope-exposure, the effects of the latter being here manifested in an unusual degree.

—C. Hart Merriam, *Results of a Biological Survey of the San Francisco Mountain Region and Desert of the Little Colorado in Arizona*, 1890

C. Hart Merriam, 1901
LIBRARY OF CONGRESS

C. Hart Merriam (1855–1942) spent his boyhood roaming the forested hills of upstate New York, observing the habits of animals and stuffing an impressive collection of specimens. Educated at home, Merriam was familiar with scientific works in the family library including Alexander von Humboldt's descriptions of plant communities found at different elevations in the Andes. Such early influences led to Merriam's lifelong fascination with biogeography: the study of why different plant and animal communities develop where they do.

Merriam's father was a U.S. congressman whose political connections led to an invitation for his son to serve as naturalist with the Hayden Survey of the northern Rockies when Merriam was just sixteen years old. Though he later trained as a medical doctor, Merriam's experience and immense enthusiasm for wildlife biology led to his appointment in 1886 as the first chief of the U.S. Department of Agriculture's Division of Economic Ornithology and Mammology (soon renamed the U.S. Biological Survey).

Merriam set about studying the distribution of wild animals and plants as a way to help farmers decide which livestock and crops to raise in different parts of the country. He sent for specimens from

OPPOSITE:
AMERICAN PIPIT: STEVE BYLAND/BIGSTOCK
DUSKY GROUSE: JON KAUFFELD/THINKSTOCK
PORCUPINE: U.S. FISH AND WILDLIFE SERVICE
ABERT'S SQUIRREL: NATIONAL PARK SERVICE
JACKRABBIT: STEVEN LOVE/THINKSTOCK
CHUCKWALLA: KOJIHIRANO/THINKSTOCK

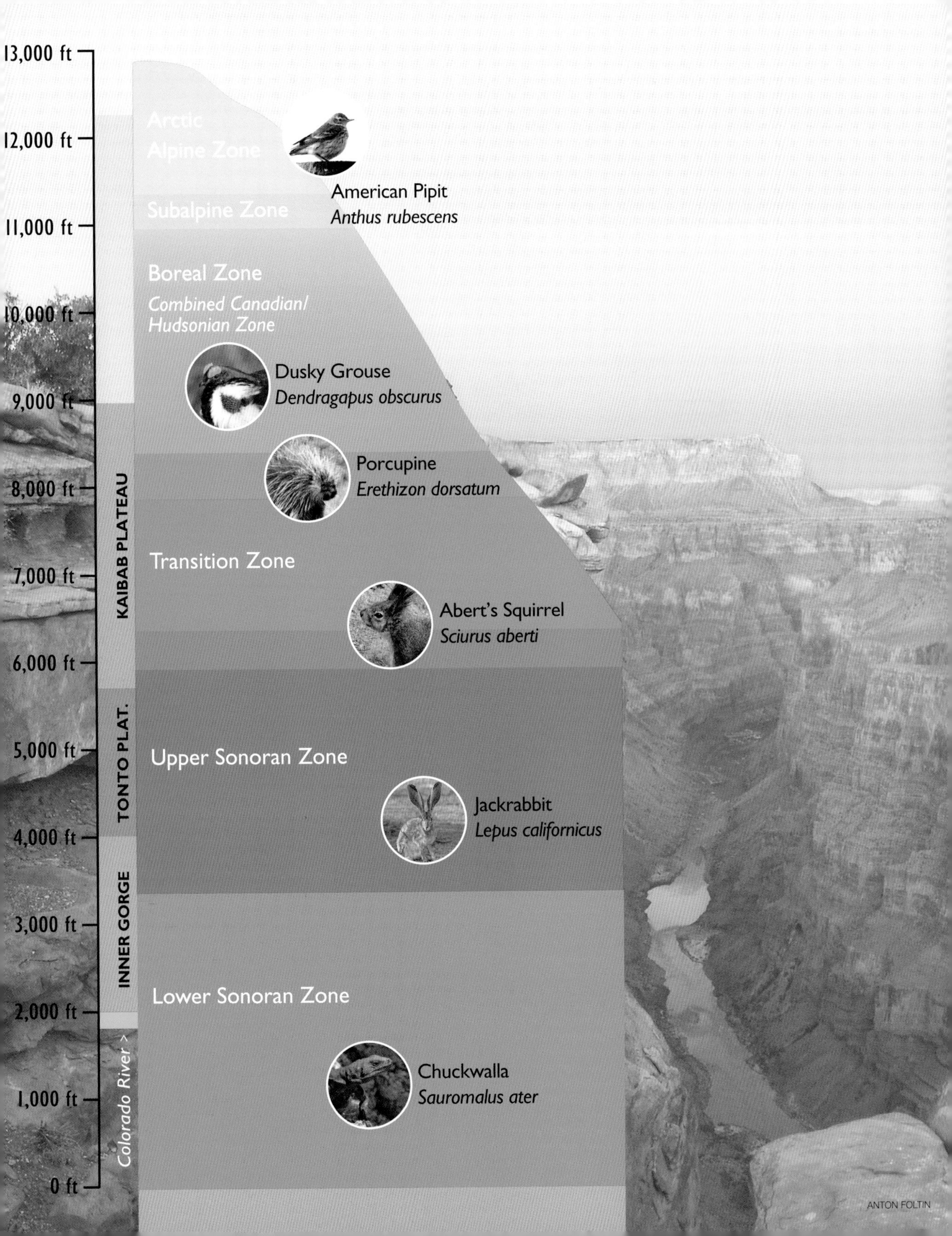

13,000 ft
12,000 ft
11,000 ft
10,000 ft
9,000 ft
8,000 ft
7,000 ft
6,000 ft
5,000 ft
4,000 ft
3,000 ft
2,000 ft
1,000 ft
0 ft
KAIBAB PLATEAU
TONTO PLAT.
INNER GORGE
Colorado River >
Arctic
Alpine Zone
American Pipit
Anthus rubescens
Subalpine Zone
Boreal Zone
Combined Canadian/
Hudsonian Zone
Dusky Grouse
Dendragapus obscurus
Porcupine
Erethizon dorsatum
Transition Zone
Abert's Squirrel
Sciurus aberti
Upper Sonoran Zone
Jackrabbit
Lepus californicus
Lower Sonoran Zone
Chuckwalla
Sauromalus ater
ANTON FOLTIN

throughout North America and plotted them on a map. In keeping with his expectations, Merriam noted that animals and plants of cold northern regions also occur on cool, high mountains far to the south. He proposed that the coldest temperature during the blooming and breeding seasons is the most important influence on where a certain animal or plant will thrive. His concept has been modified, but today's USDA cold hardiness map still reflects this basic insight.

In 1889, Merriam led an expedition to northern Arizona to prove that the boreal environment of the Canadian northwoods reaches down along high plateaus and mountains deep into the American West. He and his small team of scientists camped for two months at the base of 12,633-foot (3,850 m) San Francisco Mountain, about sixty miles south of Grand Canyon. Merriam chose the mountain because "as is well known, different climates and zones of animal and vegetable life succeed one another from base to summit."

After recording plants and animals and comparing them to those collected elsewhere, Merriam named a series of seven life zones in the area. On the highest peak he found *Arctic-Alpine* plants and an American pipit. Within the narrow treeline below, Merriam noted many *Subalpine* plants. Lower down he identified *Hudsonian*

American pipit
WOLFGANG WANDER

VERNON BAILEY (1864–1942)

M. & I took some traps & a bag of pancakes & our guns & started down in the cañon.... there is a bench of harder rock jutting out with loose rocks & soil on it & queer plants that belong away south of here. It is a hot place down there.
—Vernon Bailey in a letter to his family, 1889

C. Hart Merriam's favorite contributor of wildlife specimens was Vernon Bailey, the teenaged son of Minnesota pioneers on what was then the country's western frontier. In 1889, Merriam invited Bailey to join his expedition to northern Arizona. They had many adventures together, from hiking to the bottom of Grand Canyon to camping on top of the San Francisco Peaks, where Merriam noted: "We ate broiled eagle for supper."

Despite Bailey's youth and lack of formal education, Merriam appointed him chief field naturalist of the U.S. Biological Survey the following year. Bailey capably directed the survey's research and kept its collections organized during the agency's most exciting and productive period, from 1890 to 1933. Bailey married Merriam's sister Florence, an ornithologist and author. Their research together at Grand Canyon resulted in Vernon's 1935 *Mammals of the Grand Canyon Region* and Florence's charming 1939 *Among the Birds in the Grand Canyon Country*. Throughout his life, Bailey remained an influential voice for wildlife research and conservation on a national scale.

spruce and fir with red squirrels and dusky grouse, which he later combined with a *Canadian Zone* just below it into a single *Boreal Zone*. Down in the deserts around the mountain, he labeled two more zones the *Upper* and *Lower Sonoran* based on animals of the arid Southwest that "enter the United States from Mexico." In between the Boreal and Sonoran, he defined a *Transition Zone* of ponderosa pines with a mix of species from either direction.

Merriam wasn't quite finished. Near the end of his survey he spent a few days exploring Grand Canyon. He and his assistant Vernon Bailey marveled at the great diversity of life there and collected everything they could shoot, including a flammulated owl. "I shot a single specimen of this exceedingly rare owl while climbing out of the cañon about 3 o'clock in the morning of September 13. Its stomach contained a scorpion and the remains of insects."

Vernon Bailey and C. Hart Merriam during the U.S. Biological Survey, 1891
U.S. GEOLOGICAL SURVEY

Merriam recognized his tidy series of life zones at Grand Canyon but also observed that plants on south-facing slopes were much more sun- and heat-tolerant than those on shady north-facing slopes. He described how the canyon's rugged topography creates many microclimates due to different exposures to the sun, occasional seeps or springs of water, and soils from different rock layers, "...the effect of which on the animal and vegetable life of the cañon has been to bring into close proximity species characteristic of widely separated regions....In short, the Grand Cañon of the Colorado is a world in itself."

Air rising from the canyon's depths creates a warm, dry "rim effect."
TOM BEAN

MIXED-CONIFER FOREST and HIGH MEADOW of the KAIBAB PLATEAU: Merriam's Boreal Zone

The Kaibab Plateau's brief warm season teems with migratory birds arriving to breed and local wildlife revived by the sun.
TOM BEAN

The Kaibab Plateau rises from Grand Canyon's North Rim in a long oval mound like the back of a spoon. Here from about 8,200 to 9,200 feet (2,500 to 2,800 m) above sea level, abundant rain and snow sustain spruce and fir forests similar to the Canadian northwoods. This is the Boreal Zone, named for the Roman god of the north wind, Boreas. Smallest in extent of Grand Canyon's life zones, it is a refuge for plants and animals that can tolerate cold but need plenty of water.

Snow falls on the plateau as early as September and lingers into May, totaling up to 150 inches (4 m) in a season. Late summer rains are heavy too. The cool and soggy climate fosters dense forests where the light is dim and colors are muted.

Yet summer thunderstorms ignite wildfires that—together with tree-toppling winds and the rise and fall of insect populations—generate a dynamic, ever-changing pattern of young and old-growth trees, forest clearings, and sun-loving stands of aspen. Sweeping meadows flourish in places too challenging for any trees, especially in shallow valleys with saturated soil such as Robber's Roost, the Basin, and DeMotte Park. An especially rich mingling of creatures occurs where different habitats meet.

When snow blankets the high plateau, some animals hibernate. Many birds and mammals migrate to warmer environments. A few hardy animals remain active, however, surviving winters that can last more than seven months. Three-toed woodpeckers peck at the bark of dying conifers for beetle grubs. Dusky grouse eat needles from the ground-sweeping "skirts" of Engelmann spruce. Red squirrels feed on their hidden caches of seeds and mushrooms. Out in the meadows, long-tailed voles and northern pocket gophers gnaw on twigs and roots under the snow.

Spring arrives sometime in May, completely transforming the Boreal Zone. Fresh green foliage and ancient dark conifers alike come alive with countless insects and the singing of a rainbow of migratory birds in colorful breeding plumage. Mule deer and coyotes reappear in forest and meadow; ground squirrels and gophersnakes come out of hibernation. Broad-tailed hummingbirds newly-arrived from Mexico guard bright red, nectar-rich paintbrush. Flocks of wild turkeys feed on seeds and berries, insects, and the occasional tiger salamander. Dwarf shrews slip silently through grasses and sedges

Winding through stately vistas of the coniferous forest, quiet with pale sunlight and deep shadows…we seemed to have entered a forgotten world of deeper satisfactions.

—Florence Merriam Bailey, *Among the Birds in the Grand Canyon Country*, 1939

Mule deer
EARLE KEATLEY/THINKSTOCK

hunting voles, least chipmunks, and deer mice. The blooms of white and wanderer violets host caterpillars of the Schellbach's Atlantis fritillary butterfly, first discovered at Grand Canyon.

At forest edges, male dusky grouse perch on fallen logs with their tails fanned in a half-circle, puffing up red throat sacs and summoning females with deep booming calls across the meadows. Common nighthawk males plummet in display at dusk, making a peculiar zinging sound with their wings as they pull up just short of the ground.

Sinkhole ponds fill hollows created by water seeping through bedrock.
TOM BEAN

Aspen groves shimmer with an amazing array of little native flies and spiders, attracting energetic songbirds from warbling vireos to yellow-rumped warblers. Woodpeckers, nuthatches, and chickadees peck out nesting cavities in the soft trunks of aging trees.

Sinkholes in the forest form sunny pools ringed with aspens, conifers, and fluttering robinia with its pink pea-flowers. Flycatchers perch at the tips of branches and dragonflies cling to horsetail rush, swooping out to nab small insects buzzing above the water. When evenings grow warm enough, Great Basin spadefoot toads arise from dormancy, chorusing in grating voices to attract females to wetlands where tracks in the mud reveal how essential water is to all wildlife here.

Autumn begins earlier on the high Kaibab Plateau than elsewhere at Grand Canyon. It is a brief season, radiant with the gold and scarlet leaves of aspens, maples, oaks, and berry bushes. Summer birds and some mammals migrate south or to lower elevations. Those who remain year-round fatten on small fruits, seeds, insects, and rodents to survive the cold and snowy months ahead.

Flameskimmer dragonfly
SUSAN LAMB

THE WOODRAT CHRONICLES

Near the museum, a San Francisco Mountain woodrat has painstakingly built himself a home…in addition to dirt and sticks, it contained: 1 piece of gauze. 3 blue-headed pins from the museum. 2 burnt matches. 3 different shades of thread. 1 piece of tire tape. A quantity of electric light wiring. 1 piece of tinfoil. 1 button. 1 piece of isinglass. 1 cigar wrapper. 1 length of lantern slide binding tape. A few pinyon nuts. A quantity of juniper berries.

—Naturalist Russell Grater, *Grand Canyon Nature Notes*, October 1934

Life zones are woven so deeply into the stone walls of Grand Canyon that they seem as timeless as the canyon itself. But paleoecologists—scientists who study past environments—know that the distribution of plants and animals slowly changes over time. They have a surprising source of evidence: woodrat dens.

Woodrats collect seeds, leaves, fruits, bones, and objects such as safety pins, creating heaps of debris called middens. Under certain conditions the contents are preserved in amberat: concentrated woodrat urine rich in dissolved calcium.

Paleoecologists find Douglas-fir needles like those of today's Boreal Zone in packrat middens far below the present range of the trees. Using radiocarbon dating, they place these and other midden contents in the Pleistocene, the most recent ice age, when Grand Canyon's climate was much colder and wetter than it is today.

About twelve thousand years ago, Pleistocene conditions began to warm. Plants and animals gradually migrated higher in elevation, and usually farther north. Woodrat middens reveal that many plants now grow 2,300 to 3,000 feet (700 to 900 m) higher at Grand Canyon than during the Pleistocene, and as much as 430 miles (700 km) upstream.

The middens may also reveal which kind of packrats built them. Today five species of woodrats inhabit Grand Canyon, each adapted to specific conditions. Which of these deserves credit for the Woodrat Chronicles?

Up to seventeen inches (0.4 m) long, bushy-tailed are the largest woodrats here, with the biggest ears and bushiest tails. Studies show that they ranged farther south thousands of years ago but are now only on the north side of the Colorado River. Today they mostly inhabit rim country with ponderosas and Douglas-fir, accumulating huge piles of debris in crevices and caves. They use the same dens for generations, cementing large sections of them with amberat. The contents, locations, and architecture of their middens make it likely that bushy-tailed woodrats preserved the records of how plant communities moved as the Pleistocene ended.

NATIONAL PARK SERVICE

PONDEROSA FOREST: Merriam's Transition Zone

[The ponderosa pine] grows sturdily on all kinds of soil and rocks, and, protected by a mail of thick bark, defies frost and fire and disease alike, daring every danger in firm, calm beauty and strength.

—John Muir, *Our National Parks*, 1901

In general, the Transition Zone mixes animals from cool northern and warm southern regions of the continent.

TOM BEAN

Ponderosa pines dominate the Transition Zone on both rims of Grand Canyon. These enormous trees naturally occur in colonies of fragrant reddish-golden trunks surrounded by open grassy areas. Their upper branches mingle in canopies feathered with bundles of long, shiny needles. Slanting light reaches under them in early morning and late afternoon, while the sun's full force drenches the grasses, wildflowers, and shrubs of the clearings all day. The combination of "tree islands" and open areas offers many possibilities for forage and shelter.

Clumps of ponderosas are like ancient villages. Their occupants hunt and gather nourishment, seek cover, and raise their young at different levels from beneath the soil to the tops of the trees to out in the grassy openings. Goshawks and squirrels, mountain lions and mule deer, lizards and tarantulas—all have their places in an ecosystem adapted to frequent low-intensity fires and strong but erratic seasonal changes.

Ecologists describe several ponderosa pine associations where ponderosas grow with a specific grass, shrub, or tree. The ponderosa/Gambel oak association is an important example. Gambel oaks do best in poor, stony soils and yet are hot spots of biodiversity. They host many different insects, including several butterflies. Wild turkeys and mule deer relish their acorns in autumn. Mice, squirrels, and acorn and Lewis's woodpeckers store them for later.

Both ponderosa pines and Gambel oaks naturally form clumps that foster unique wildlife communities.

TOM BEAN

Life zones below the rim are too hot and dry for most Transition Zone animals to pass through them from one side of Grand Canyon to the other. Transition Zone animals confined to the north side include yellow-haired porcupines, bushy-tailed woodrats, mountain cottontails, northern pocket gophers, and least and Uinta chipmunks. Among those found only on the south side are Arizona porcupines, desert cottontails, and Mexican voles. Charcoal-gray Kaibab squirrels, a population isolated on the North Rim, are a subspecies of the rusty-brown tassel-eared squirrels that live in ponderosa forests throughout the interior West.

Kaibab squirrel
NATIONAL PARK SERVICE

The Transition Zone's yearly precipitation of eighteen to twenty-two inches (0.5–0.6 m) is about evenly divided between winter snow and late summer rain. Winter is never as long and deep here as it is in the higher Boreal Zone, and so the Transition Zone is livelier in winter. Tracks reveal the ramblings of mountain lions, coyotes, bobcats, gray fox, mule deer, cottontails, and tassel-eared squirrels. Steller's jays squawk and swoop among the trees; mountain chickadees join mixed flocks of small birds. One hundred or more pygmy nuthatches roost together in tree cavities to keep warm.

Spring brings weeks of back-and-forth changes in the weather. Mild days take turns with raw winds and horizontal snow as the hours of daylight increase. As early as March, small flies visit ground-hugging flowers; migratory birds begin to return. Young chipmunks and ground squirrels emerge from burrows to play and bask in the sun, chipmunks with stripes across their faces and ground squirrels with stripes that stop at their necks. Short-horned lizards lap up ants, which slowly digest in their tanklike bellies.

The rare Grand Canyon ringlet butterfly depends on grasses along the South Rim for much of its life cycle.
TOM BEAN COURTESY OF LARRY STEVENS

May and June increasingly fill with life but are very dry months that especially challenge mothers and young. Ponds dry up; seeps and springs are few. Animals survive on the scant moisture they get from plants or prey.

Late summer thunderstorms bring welcome relief and a second surge of life to the Transition Zone. Bigger, brighter flowering plants and their insect partners burst upon the scene. Lightning ignites small fires that crawl through the forest, restoring nutrients to the soil and stimulating future production of the leaves, blossoms, and seeds that form the foundation of the food chain.

Autumn takes over the Transition Zone gradually. Grasses dry to a tawny color; birds eat their seeds and rodents store them. Mule deer, coyotes, and black bears gorge on the fruit of shrubs.

Coyote
TOM BEAN

PINYON-JUNIPER WOODLAND: Merriam's Upper Sonoran Zone

Continuous exposure of the north wall to the desert sun and currents of hot air arising from the canyon bottom favor the development there of an Upper Sonoran flora. The result, insofar as it affects the butterflies with their natural tendency to fly upward, is that Upper Sonoran species mingle with Canadian species along the narrow marginal strip of the canyon's North Rim.

—John S. Garth, *Butterflies of Grand Canyon National Park*, 1950

Kaibab indra swallowtail butterflies—known only from the north side of Grand Canyon—ride currents of warm air up south-facing walls to the North Rim, pausing at muddy seeps to ingest minerals for the pheromones they use in courtship. A fortunate hiker on the North Kaibab Trail may happen to see these rare butterflies, which depend on different life zones at different times of year.

Most Grand Canyon visitors get to know the Upper Sonoran Zone on paths along the rims, where updrafts from the hot and dry inner canyon foster gnarled junipers and pinyon pines that are seldom more than thirty feet (9 m) tall. These wizened trees dominate a rich life zone that spills over 3,000 feet (900 m) down canyon walls in sweeps and patches of pinyon-juniper woodland, mountain shrubs, chaparral, and sagebrush grasslands. Although Merriam named this life zone for the warm Sonoran Desert region to the south, several characteristic plants including pinyon pine, juniper, and sagebrush are actually native to the cold Great Basin desert to the north and west.

Seasons are subtle in the Upper Sonoran, where rain and snow average a scant fifteen inches (0.4 m) a year. Overall warmer temperatures result in more reptiles—lizards, skinks, and snakes—than at higher elevations. Rodents are very common, and many migratory birds spend the mild fall and winter here.

In spring and again in late summer, storms prompt two bursts of growth and blooming. Along with surges of foliage and flowers come hordes of insects and spiders pursued by lizards, birds, and bats. Leaves, flowers, and seeds sustain mice, woodrats, rock squirrels, and chipmunks. These toothy foragers are

Kaibab indra swallowtail
TOM BEAN COURTESY OF LARRY STEVENS

Pinyon pines (right) and junipers (left) are the characteristic trees of the Upper Sonoran Zone.
TOM BEAN

preyed on in turn by gophersnakes, gray foxes, coyotes, bobcats, owls, kestrels, and the occasional long-tailed weasel. Much of the action—courtship and mating, foraging and hunting—takes place at night.

Pinyon-juniper woodland is the most easily recognized plant community in this life zone. From the rim to the bottom of the Redwall cliffs, hikers on the sunny South Kaibab Trail pass through pinyon pines and juniper trees dotting a landscape open to the sky. Pinyon nuts and plump cones called juniper berries anchor the food chain, vital to pinyon and scrub jays, pinyon mice, and other wildlife. Native only to Grand Canyon, mushroom-colored scrub wood-nymph butterflies flutter among the trees, the two pairs of eyespots on their wings winking in the sun.

Conditions are more shady and cool in side canyons and north-facing alcoves along the rims. Stands of Douglas-fir—trees associated with the Pacific Northwest—have persisted since the last ice age in dim pockets where winter snows linger longest.

Scrub wood-nymph
TOM BEAN COURTESY OF LARRY STEVENS

Leafy deciduous trees and shrubs flourish on brighter bluffs and slopes, among them Gambel oak, serviceberry, snowberry, and chokecherry. Evergreen, fire-adapted chaparral develops in exposed places where sun, slope, and soil reduce the influence of summer rains. Both of these shrubby plant communities yield lots of fruits. Rodents and shrubland birds such as spotted towhees scuff in them for berries, seeds, and insects.

Where the Upper Sonoran Zone extends back away from the rims, including at Desert View, Pasture Wash, and the Toroweap Valley, there is deeper soil and flatter terrain suited to American badgers, pocket gophers, jackrabbits, and skunks. There are also some almost pure stands of sagebrush with associated animals such as northern sagebrush lizards.

TOM BEAN

Caves present interesting and odd evolutionary forms.

—Dr. Neil Cobb, Colorado Plateau Museum of Arthropod Biodiversity

Researchers have recorded over four hundred caves within Grand Canyon National Park. Members of the park's Science and Resource Management Division have found bits of condor bone dating back more than ten thousand years in some of them, as well as the remains of other now-extinct species such as the Harrington's mountain goat, Shasta ground sloth, camel, horse, and dire wolf.

These days, several of the canyon's more than twenty species of bats roost in its caves during the day, and some use them as maternity colonies. California condors and Mexican spotted owls nest and raise their young in the mouths of caves in narrow side canyons and sheer cliffs of the Redwall Limestone, where they are less vulnerable to predators.

NATIONAL PARK SERVICE

Deep, undisturbed caves are a world unto themselves, with relatively constant temperature and humidity fostering fragile communities of algae, fungi, and rare insects and spiders. Among them are millipedes, booklice, pill bugs, fungus gnats, harvestmen, scavenging beetles, and the Grand Canyon pseudoscorpion. Some darkness-adapted invertebrates are ghostly looking because they lack pigment, or have reduced or missing eyes, or elongated appendages and antennae. Harsh outside conditions prevent many of the creatures found in different caves from intermixing, resulting in caves with unique species.

Cave ecosystems are delicate and highly sensitive to disturbance and contamination. However most of Grand Canyon's caves are very remote or extremely difficult and dangerous for humans to reach. All are strictly off-limits to visitors except for the intriguing Cave of the Domes on Horseshoe Mesa.

TREELESS DESERT SCRUB: Merriam's Lower Sonoran Zone

And then, as we look lower and lower down in the shadowy depths of the canyon, we see the end of the treeline, where the desert shrubs and cactus of Mexico meet our eyes: the Lower Sonoran Life Zone.

—Ranger Chester R. Markley, *Nature Notes*, January 1931

Black-throated sparrows have been called "the avian spirit soul of the Tonto blackbrush." They are among the very few birds that breed in the stark desert scrub of the Grand Canyon's sun-blasted Tonto Platform.

Desert scrub develops from around 4,000 feet (1,200 m)—just below the steep cliffs of the Redwall Formation—to the bottom of the canyon. Rainfall averages seven inches (18 cm) a year, and summer temperatures often exceed 100 degrees Fahrenheit (38°C).

The Lower Sonoran Zone is a desert realm of rodents and reptiles.
TOM BEAN

Most visitors experience the Lower Sonoran life zone on the Tonto, a broad ledge that ends abruptly at the brink of the dark Inner Gorge. Much of the year, the Tonto Platform appears to be a monotonous stretch of nothing but scrubby little blackbrush as far as the eye can see. But in spring, the Tonto buzzes with insects pollinating cactus blossoms, fendlerbush, and creamy-white yucca flowers. Shiny black carpenter bees visit flowering agave spikes up to fifteen feet (4.6 m) tall.

Apart from the many different birds attracted to spring's brief period of flowering and insects, the low diversity of animals here reflects the Tonto's limited number of plant species, which provide little variation in nourishment or types of cover. Only a few creatures can cope with this forbidding habitat. Among the few exceptions are Paiute giant-skipper butterflies unique

Diversity is low on the Tonto Platform, and the Inner Gorge offers scant habitat for wildlife.
TOM BEAN

to the Grand Canyon, which are surprisingly common around the edge of the Tonto in August. Their caterpillars subsist in the sword-like leaves of Utah agaves, an environment that ecologist Larry Stevens describes as "like living in hot soap."

In Grand Canyon's western half, the Lower Sonoran Zone covers terrain more varied than the Tonto Platform, with greater diversity of wildlife. Animals associated with the Mohave Desert occupy pockets of desert scrub and grassland, creosotebush, and Joshua trees here.

Cactus wrens nest in the western canyon, usually in Joshua trees or large cholla cactus. On the ground, coveys of Gambel's quail rummage for seeds and greater roadrunners hunt scorpions, snakes, and lizards. Costa's hummingbirds spend early summer in this arid, low-elevation zone, then leave when the temperature rises. Mammals cope with the heat in various ways. Canyon and cactus mice forage at night and sleep in burrows during the day. Black-tailed jackrabbits doze sheltered under bushes from the sun, emerging at dusk. Coyotes, bobcats, and gray fox adapt to the moment, usually resting when it's hot and hunting when it's cooler. Like rock squirrels, white-tailed antelope squirrels are active in daytime but often pause in the shadows of rocks or plants. They climb shrubs—including cactus—to collect fruits and seeds, then curl their tails over their backs to shade themselves as they dash to their burrows to eat.

One particular group of mammals thrives here. Pocket mice and kangaroo rats are a distinctive family of rodents called Heteromyids that have fur-lined pockets in their cheeks for collecting food and long hind legs with big feet for jumping. Their lengthy tails help them keep their balance; kangaroo rats can jump as far as nine feet (3 m) in one leap. Heteromyids are nocturnal and have large heads on tiny bodies, bright buggy eyes, and big ears that can hear the swoop of an owl or the slither of a snake. They plug the openings of their burrows while they sleep to keep out the heat of the day.

Desert tortoise
NATIONAL PARK SERVICE

Reptiles in general are well-adapted to desert conditions. Desert tortoises can live for months without water, obtaining moisture from plants. They have a slow metabolism, hibernating for several months in winter and spending much of their long lives burrowed underground, where it is cooler than on the surface.

At least a dozen different lizards and eighteen species of snakes thrive in the Lower Sonoran. Rattlesnakes—including the rare Grand Canyon pink—hunt lizards and small mammals mostly at night, using heat-sensitive pits between their eyes and noses to sense their warm-blooded prey.

EDWIN DINWIDDIE MCKEE (1906–1984)

In the middle of this white outcrop, coiled like a rope, and just as motionless, was a medium-sized salmon-pink rattlesnake.…It seemed clearly up to me, as Park Naturalist at Grand Canyon, to learn more about this specimen; it required collecting the snake alive in order not to injure the diagnostic scale pattern of its head.…On top, my car was waiting among the trees and I had only to drive a few miles over a winding dirt road to the Desert View Ranger Station.…I had to continue holding the snake in one hand while I drove with the other. I recall clearly that I was carrying the pink rattler in my right hand and reasoned that should I drop it for any reason while driving, it would be near my feet. Thus a transfer to my left hand had to be arranged. Finally, with snake dangling out of the window, I drove in low gear to the station and arrived with my specimen conspicuously on display, much to the amazement of the local ranger.

—Edwin D. McKee, *Journal of Arizona History*, 1976

Although he grew up in Washington, D.C., Eddie McKee heard many stories of the Grand Canyon as a boy. His scoutmaster was François Matthes, a famous cartographer who in 1903 had made the first small-scale topographic maps of the canyon. In 1927, Matthes arranged a summer internship for McKee with the Carnegie Institution. McKee was assigned to help develop an educational program at the Grand Canyon for the National Park Service.

NATIONAL PARK SERVICE

McKee's experience at the canyon changed his life. He enrolled at Cornell University to study geology and spent summers working in the national park. When park naturalist Glen Sturdevant drowned while attempting to cross the Colorado River in 1929, Eddie McKee was hired to take his place.

McKee continued to explore and record the canyon's resources. On a hike down the Tanner Trail to "do some geological work," he made his famous discovery of the Grand Canyon rattlesnake. McKee made many other hikes across the canyon that same year to visit Miss Barbara Hastings, who was working on the North Rim with Florence and Vernon Bailey for the U.S. Biological Survey. Eddie and Barbara were married that December.

McKee was a well-rounded naturalist of the old school who collected butterflies and created checklists of the park's mammals, amphibians, reptiles, and birds. Having been formally trained in geology, he published many scientific papers on Grand Canyon sediments and stratigraphy. He and Barbara ran a bird-banding station together, collected Havasupai baskets, and socialized with the scientists, park staff, and indigenous people they met at Grand Canyon. McKee named features and formations of the Supai Group after their Havasupai friends.

Eddie McKee left the Grand Canyon in 1938. He continued his career as a geologist, traveling to every continent except Antarctica to study and publish his findings about sedimentary geology. He and Barbara now rest in Grand Canyon Cemetery by a simple marker of Tapeats Sandstone.

RIPARIAN

Springs ecosystems are among the most ecologically complex, biologically diverse, culturally valued, and threatened ecosystems on Earth.

—Dr. Larry Stevens, Curator of Ecology, Museum of Northern Arizona

We can sense seeps and springs before we see them. The scent of muddy soil, a tinge of humidity in the air, the chirps of a fly-catching phoebe let us know an oasis is near.

Riparian habitat (from the Latin *ripa*, "riverbank") develops alongside flowing water, around ponds and lakes, and where groundwater emerges as a seep or spring. Above 5,000 feet (1,500 m), riparian areas can form a gallery forest of alder, willow, boxelder, maple, and ash trees along with shrubs, sedges, and other flowering plants. At lower elevations they often sustain willow, cottonwood, and introduced tamarisk. Seeps and some springs may be too small to support trees but still develop a strikingly lush community of plants and insects compared to the dry life zones around them.

At least forty-eight species of birds nest in Grand Canyon's riparian habitats. Bark scorpions and countless insects are linked to it at some stage in their life cycle. Bighorn sheep, raccoons, skunks, several bats, several lizards and snakes, canyon tree frogs, and Woodhouse's and red-spotted toads all require reliable access to water. Mammals and birds from miles away depend on riparian moisture.

The riparian habitat itself is really a series of microhabitats radiating out from a source of water: marshy ground with horsetail rushes and dragonflies, damp soil with sedges and willows, floodplain with arrowweed and scorpions, and places where cottonwoods sink their roots into moist soil. Waterfalls support algae and snails on the rocks they flow over, columbine and bumblebees in the spatter zones on either side, and maidenhair fern and vireos around the pools beneath them. In alcoves hung with moss and monkeyflowers, overhanging seeps drip into tiny pools visited by an array of birds that wait in turn to drink.

Surface water is rare at Grand Canyon yet essential to many forms of life.
TOM BEAN

Springs cover less than .01 percent of Grand Canyon's area but support up to five hundred times as many species as their surroundings. They create habitats that vary from hanging gardens to "gushets" that spout from canyon walls, as at Thunder River. Due to their individuality and isolation, springs foster many rare and unique species including the Vulcan's Well giant waterbug, and the wetsalts tiger beetle and masked clubskimmer dragonfly found only between Stone and Nankoweap Creeks.

AQUATIC: THE COLORADO RIVER

The Colorado is never a clear stream, but for the past three or four days it has been raining much of the time, and the floods poured over the walls have brought down great quantities of mud, making it exceedingly turbid now.

—John Wesley Powell, *The Exploration of the Colorado River and Its Canyons*, 1895

Colorado is one of several words for "red" in Spanish. The Colorado River was named for the rusty-red sediments it once carried year-round.

When John Wesley Powell led the first known trip through the Grand Canyon in 1869—with a handful of men in wooden boats—the Colorado was a warm, silty river that ran high and furious with snowmelt in spring and again with runoff from late summer rainstorms. It sustained river otters, beavers, muskrats, and several peculiar native fish adapted to seasonal ups and downs in the river's temperature, volume, and sediments.

Exploration and settlement of the West brought major changes. Nineteenth-century trappers eliminated most of the beavers, otters, and muskrats. Introduced tamarisk trees have spread along the Colorado and its tributaries, crowding out native flora including willows and making the riverbank soil more salty. Completion of Glen Canyon Dam upstream in 1963 created Lake Powell, where most of the sediment that once flowed into the Grand Canyon now settles. Because water released from the dam comes from deep in the reservoir, the river is now cold as well as clear, resulting in poor conditions for the spawning of native fish.

Three native fish—bonytail chub, roundtail chub, and Colorado pikeminnow—have been eliminated from the river within Grand Canyon. The native humpback chub is endangered. Introduced carp, fathead minnows, and rainbow, brown, and brook trout are common; they prey on native fish, especially their young, and compete with them for food.

Yet the Colorado River remains an astonishingly rich ecosystem. The ecology of the river and its tributaries is complex. It begins with diatoms (single-celled algae with silicate cell walls) and zooplankton, and moves up through aquatic invertebrates and the larvae of caddis flies, mayflies, dragonflies, and stoneflies. There are over thirty crustaceans, about half of which are microscopic. These little river-born creatures are consumed by fish, canyon tree frogs, red-spotted and Woodhouse's toads and in many cases, by each other. Three native fish—the Colorado River speckled dace, flannelmouth sucker, and bluehead sucker—are still common. The dynamics of the Little Colorado River—the Colorado's main tributary within Grand Canyon—are intact, making it possible for endangered humpback chub to reproduce there. In 2012, researchers caught and released the first razorback sucker to have been found in the river since 1990.

The river corridor continues to provide safe passage and shelter for more than two hundred migratory seafowl and wintering birds. Observers have recorded nineteen species of waterfowl wintering on the river between Lees Ferry and Soap Creek. In 2000, biologists and volunteers surveying the length of the Colorado River within the park found almost three hundred beaver dens as well as the tracks of a few muskrats and an otter.

Today's resource managers have access to extensive scientific research to inform their decision making relative to the Colorado River.

TOM BEAN

EPILOGUE: CREEKS

You can never step into the same river; for new waters are always flowing on to you.

—Heraclitus of Ephesus, 540–480 BC

Nankoweap, Bright Angel, Deer, Shinumo, Havasu, Diamond... the names of these ever-flowing creeks bring to mind much that is marvelous about the Grand Canyon: its ancient peoples, its adventurous explorers, its mind-bending geology, and its many different and fascinating animals.

They also remind us that nature is dynamic, ever-changing. The Grand Canyon's life zones are useful generalizations, but even a casual observer can see that there are no sharp boundaries between them. Their edges are blurry; their plants and animals intermingle. Such movement and mixing is essential because it allows for what scientists call "gene flow and species dispersal," which keep populations vigorous and adaptable to change, whether it is temporary or long-term.

Bright Angel Creek is a vital source of life in the heart of the Grand Canyon.
TOM BEAN

The bright streams that flow down through the Grand Canyon's layers lace its life zones together, providing a passageway as well as a lifeline for present and future forms of life. They remind us that the Grand Canyon is about so much more than rocks; it is a vast refuge for living things, whether plant or animal, wild or human.

GRAND CANYON WILDLIFE: SELECTED SPECIES

The following list includes animals that are common—for example, rock pocket mice—or are commonly seen, such as mule deer, rock squirrels, and ravens. Certain animals are on the list because visitors often ask about them, especially black bears and mountain lions. Several species are unique to Grand Canyon, typical of the Southwest, indicators of a healthy ecosystem, or of concern to park management because they are threatened or endangered.

Mammals

Bat, spotted *Euderma maculatum.* Nocturnal. Summer: coniferous forests. Winter: Upper Sonoran, Lower Sonoran, riparian.

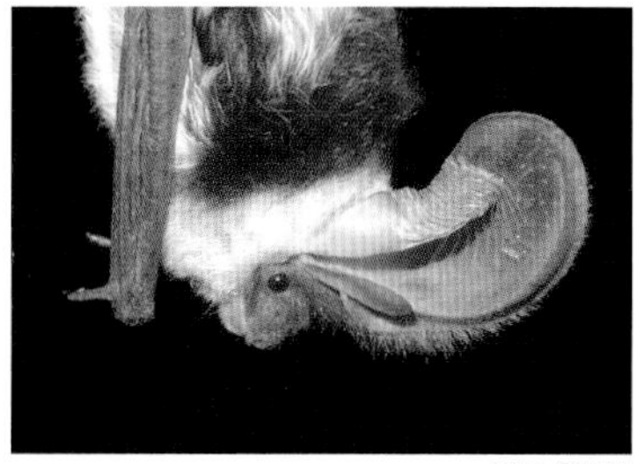

PAUL CYRAN

Bats are little furry flying mammals more closely related to primates like human beings than to rodents. Grand Canyon has twenty-two species of bats from three families, each species with its own habitat, feeding and breeding behavior, and daily schedule. Among the most likely to be seen are spotted bats. Migrating seasonally between all habitats at the Grand Canyon, they have three big striking white spots on their backs and long webby ears that look almost like another set of wings. They emit high metallic shrieks audible to humans as they hunt for noctuid moths.

Bear, black *Ursus americanus.* Mostly crepuscular: rolling landscapes of mixed conifer, ponderosa, pinyon-juniper.

KEN HOEHN/THINKSTOCK

Black bears lumber flat-footed through forests, grazing on grubs, leaves, berries, and very occasionally gnawing on carrion or pouncing on small animals. They doze in dens during the winter, occasionally emerging on fine days. One to five cubs are born while their mother sleeps.

Beaver, American *Castor canadensis.* Crepuscular/nocturnal: aquatic.

JOHN PITCHER/THINKSTOCK

Largest of North American rodents, beavers den in the banks of the Colorado River and build dams on its tributaries that often wash out in spring floods. Playful and affectionate, they mate for life. Both parents care for their kits with the help of older siblings. They eat aquatic plants, shoreline grasses, and the inner bark of branches, and can close their mouths behind their front teeth to carry sticks without swallowing water. Designed for swimming, they have flat, scaly tails and webbed hind feet with claws for grooming their dense fur, which on Grand Canyon beavers is more cinnamon than brown.

RON SANFORD/THINKSTOCK

Bobcat *Lynx rufus*. Mostly nocturnal: all habitats except aquatic.

Lanky and long-legged, bobcats have reddish-tawny coats with black spots, tufted cheeks with curving black stripes, and short tails that end in a stub. They lie in wait to ambush prey and can bring down animals much larger than themselves with a suffocating bite. Not only do they look much like large house cats, they also meow, purr, and screech like them.

MU_MU/THINKSTOCK

Chipmunk, cliff *Tamias dorsalis*. Diurnal: rocky cliffs, pinyon-juniper.

Chipmunks eat leaves, buds, flowers, seeds, and mushrooms and store some for later, as well as insects, eggs, and occasional dead things. Hyperactive during the day, they rest in burrows at night and through the winter, awakening to eat now and then. Park naturalist Glen Sturdevant recorded a cliff chipmunk's monotonous barking for three separate minute-long intervals one afternoon in 1928: "His barks numbered 172, 146, and 162, respectively...each bark was accompanied by a twitch of the tail."

Coyote *Canis latrans*. Active at all hours: all habitats.

Coyotes will eat almost anything—juniper berries, chokecherry seeds, rabbit fur, beetle husks—any time, any place. Lone coyotes patrol urine-marked hunting trails but also hunt together, chasing a deer in relays until it tires or crossing a meadow snapping up grass-hoppers flushed by one another's paws. On pleasant evenings coyotes howl in a chorus converging from all directions. They are sociable and playful and usually mate for life.

Deer, mule *Odocoileus hemionus*. Active at all hours: all terrestrial habitats (see page 55).

Numerous and social, mule deer forage together on shrubs, grasses, and leafy plants in summer and twigs and evergreen needles

NATIONAL PARK SERVICE

in winter. Ever alert to threats, they may suddenly bolt and bounce in stiff-legged hops called stotting or pronking, sometimes pausing to look back. Males grow antlers that rise at a high angle and fork in pairs into a wide crown. They lose these in late winter but soon begin regrowing them, the new antlers encased in fuzzy, blood-rich skin called velvet. In fall bucks rub off the velvet and their necks swell. They rut in midwinter, snorting, roaring, grunting, and smashing their antlers into each other. The winner mates with the doe and she gives birth seven months later to one or two gangly, spotted fawns that bleat as they follow her in the forest.

BRENT RIEGSECKER/THINKSTOCK

Elk, Rocky Mountain *Cervus elaphus nelsoni*. Active at all hours: ponderosa forest, Upper Sonoran. South Rim only.

MIKE BUCHHEIT

Rocky Mountain elk are not native to the Grand Canyon area. Merriam's elk, a native subspecies, ranged between 8,000 and 10,000 feet (2,400 and 3,000 m) in Arizona's lush White Mountains until they were hunted to extinction in the early twentieth century. Arizona Game and Fish Department transplants of non-native elk to northern Arizona began in 1913 with several dozen Rocky Mountain elk from Yellowstone National Park. Elk depend heavily on fresh water, and artificially supplemented water sources have allowed them to greatly expand their range since introduction. They reached the South Rim of Grand Canyon in the 1970s and now occur in great numbers due to abundant water from stock tanks. The absence of predators in developed areas, especially Grand Canyon Village, enables elk to flourish there.

Fox, gray *Urocyon cinereoargenteus*. Active at all hours: all habitats.

Gray foxes are small members of Canidae—the dog family—only about eighteen inches (0.5 m) long but with magnificent bushy tails a foot (0.3 m) or so in length. Coarse white hairs on their reddish sides and black backs make them look grizzled. Semiretractable claws enable them to climb trees to rest and hide, the only members of Canidae to do so. They forage for mice, cottontails, squirrels, birds, insects, berries, cactus fruit, and grasses. Pairs stay together with

RON SANFORD/THINKSTOCK

their offspring for up to a year. Gray foxes communicate with mews and murmurs when close together. They growl and snarl in confrontations and bark and scream when far apart, but never howl like coyotes.

FUSE/THINKSTOCK

Lion, mountain *Puma concolor*. Crepuscular/nocturnal: mostly rim country.

Mountain lions are strict carnivores with powerful jaws that have bladelike carnassial molars for carving up large herbivores. At Grand Canyon they haunt forests along the rims, watching from a ledge or rock outcropping or thicket until a mule deer or an elk browses close enough for them to spring from hiding and seize it. Male lions mark their home range with scrapes, pushing up and often urinating on small piles of pine needles or other debris. Their ranges vary from twenty-five to hundreds of square miles, depending on the abundance of prey, the kind of vegetation, and the season of the year. Mountain lions avoid each other mostly by scent, scrunching their faces in a flehman grimace to inhale pheromones into the vomeronasal organ on the roof of their mouths. Females ready to mate, however, find males this way. Mother lions nurse their young for eight weeks or more, delicately grooming them with tongues capable of stripping fur from kills.

B. MOOSE PETERSON

Mouse, grasshopper *Onychomys leucogaster*. Nocturnal: Upper Sonoran.

Mice have poor eyesight but keen hearing, an acute sense of smell, and squeaky voices. Baby mice cheep when cold, and males of some species sing courtship songs in a range too high for humans to hear. Although seldom seen by humans, mice are a major source of nourishment for predatory mammals, birds, and snakes. Most mice feed on green plants, seeds, fruits, and insects, but the northern grasshopper mouse is a fierce carnivore that, when challenged, lifts its tiny head and howls.

Pocket mouse, rock *Chaetodipus intermedius.* Nocturnal: inner canyon south of Colorado River (see page 55).

TATIANA GETTELMAN

Pocket mice and kangaroo rats have long hind legs and feet for jumping, putting them in the category of Heteromyids, or jumping mice. Kangaroo rats are uncommon at Grand Canyon, but there are two common pocket mice here: Great Basin pocket mice north of the Colorado River and rock pocket mice south of the river in rocky areas from the Tonto Platform to the South Rim.

Porcupine, North American *Erethizon dorsatum.* Crepuscular/nocturnal: mostly rim forests.

U.S. FISH AND WILDLIFE SERVICE

The Colorado River separates two subspecies of porcupines, the yellow-haired to the north, with greenish-yellow tips on long hairs, and Arizona porcupines on the south that are smaller, with white-tipped hair and bigger ears. Porcupines patrol south-facing slopes and ridges in coniferous forests. By day they sleep among rocks or high in trees; at night they feed on plants and the inner bark of twigs, and swallow mud and rocks for minerals. They react strongly to footsteps and flip their tails when alarmed. Bowlegged with long claws and weak eyes, they feel their way with sensitive whiskers, guard hairs, and a keen sense of smell. They smack when they eat and snuffle, growl, boom, chatter, squeak, and bark—to the alarm of humans nearby. Baby porcupines sound like kazoos. Vernon Bailey wrote that when excited, his pet porcupine whirled around and hopped up and down "like an elephant might be supposed to dance."

Pronghorn *Antilocapra americana.* Diurnal: open Upper Sonoran on South Rim.

Pronghorns are the only remaining species of Antilocapridae, an ice age family of mammals. They have enormous, far-seeing eyes and favor open landscapes where they can easily spot and flee from predators. They have huge hearts and lungs that enable them to run long distances. The swiftest mammals in North America, their top speed has been clocked at fifty-five miles (90 km) an hour. Pronghorns shed the black sheath from the bony core of their forked horns each year.

U.S. FISH AND WILDLIFE SERVICE

ALAN VERNON

Rabbit, desert cottontail *Sylvilagus audubonii.* Crepuscular: shrubby habitats with open areas.

Cottontails emerge at dawn and dusk from sheltering shrubs or logs to browse on plants, paying close attention to shadows passing overhead and subtle sounds warning them of red-tailed hawks, owls, coyotes, snakes, and myriad other predators. With such odds against their survival, they reproduce often. Nuttall's cottontails are a subspecies found on the North Rim and Shiva Temple in sagebrush or where trees meet meadows. Desert cottontails are common south of the Colorado River.

TATIANA GETTELMAN

Ringtail *Bassaricus astutus.* Nocturnal: all rocky habitats (see page 55).

Ringtails are members of the raccoon family that look a bit like cats but have pointed muzzles, large ears, and huge furry tails ringed with black and white stripes. Their bulging eyes see well in the dark as they silently slip among rocks to hunt insects, lizards, mice, and berries. They can cartwheel to reverse direction or climb trees to raid bird nests. Their rotating ankles allow them to "chimney" up narrow chutes. Otherwise solitary and territorial, ringtails pair and remain together after mating until their young are weaned. Ringtails have a loud bark, and click and chatter like their raccoon cousins.

NATIONAL PARK SERVICE

Sheep, desert bighorn *Ovis canadensis nelsoni.* Diurnal: rocky inner canyon (see page 55).

Unlike most mammals, desert bighorns can tolerate fluctuations of several degrees in their body temperature. They are able to survive for weeks without water during the nonsummer months and quickly recover from dehydration. During fall and winter, bands of twenty or fewer bighorn sheep frequent the upper walls

TERRITORIES OF TIME

For humans there is no wilderness more difficult to explore than the night....We can only speculate on the range of color and tonality perceived by eyes adapted to the light of moon and stars.

—David Rains Wallace, *The Dark Range: A Naturalist's Night Notebook*, 1978

Although C. Hart Merriam explained where we might see certain animals, he didn't say *when*. Grand Canyon is a refuge for at least 350 birds, ninety-five mammals, fifty-four reptiles, twenty-nine fish, eight amphibians, and countless insects and spiders. Yet even a lucky visitor will probably see only ten or twenty of these on a day's visit to the park. Where is everybody?

Animals not only inhabit a place, they also inhabit a time. They are active during hours when they can best avoid predators, forage for food, find or create shelter, mate, and care for young—all while striking a balance between their need for warmth and their need for water, between the calories they consume and the energy they expend.

Most of us seldom see wild animals because so much animal activity takes place at night. *Nocturnal* animals tend to have more acute senses of hearing and smell as well as eyes that gather more light than we humans have. At night they are safe from the sun's drying heat and damaging radiation as well as from certain predators. They also avoid competing for food with animals that forage during the day. Badgers and bats, moths and shrews, mice and rats, raccoons and ringtails, weasels and mountain lions—are all mainly nocturnal, as are most owls and amphibians and many snakes.

Diurnal animals are active during the day: squirrels, chipmunks, prairie dogs, butterflies, some reptiles, and most birds. Warmed by the sun, they spend their energy foraging for plants and insects to eat, courting mates with coloring and displays, and staying alert for the movements and shadows of both prey and predator.

Crepuscular animals (from the Latin *crepusculum*, meaning twilight) move about in the dim hours around dawn and/or dusk, when they are difficult for predators to see and not stressed by the heat of day. Mule deer, elk, rabbits, and skunks are among the animals considered to be crepuscular.

MARTIN VALLIÈRE/THINKSTOCK

As we continue to collect and share information about the behavior of our fellow animals, the categories of nocturnal, diurnal, and crepuscular are yielding to a more flexible understanding of the natural world. Some animals—coyotes, for example—feed or rest at any time of day depending on their need of the moment. Young black bears play and explore in the daytime; adult bears are more crepuscular. Other animals may modify their patterns of rest and activity depending on seasonal temperatures, the increase or decline of a certain kind of food plant or prey animal, or even the phases of the moon.

of the Grand Canyon, nibbling plants in the mornings and late afternoons. Males joust for dominance by rearing up on their hind legs, running a few steps, and bashing each other with their heavily ridged horns. In spring, males migrate to join females at lower elevations for the lambing season. Even more nimble than their muscular parents, the lambs leap and play and conduct mock battles on precipitous ledges. Their mothers stop nursing them in response to diminishing levels of protein in their food plants.

NATIONAL PARK SERVICE

Shrew, Crawford's gray *Notiosorex crawfordi*. Nocturnal: Upper and Lower Sonoran.

Shrews have tiny eyes, orange or dark red teeth, and minuscule bodies that lose heat rapidly, requiring them to eat three times their weight daily in short, frenetic bouts of foraging alternating with deep naps. They consume very small prey, especially crickets, beetles, spiders, and larvae. Shrews have long pointed snouts and an unpleasant smell from glands on their flanks. It is said they are so ferocious and territorial that even mating is a struggle against the urge to kill. Dwarf and Merriam's shrews live in Grand Canyon's high-elevation forests and meadows. Crawford's gray shrews occupy drier areas and obtain the water they need from their food.

Skunk, spotted *Mephitis gracilis*. Nocturnal: Lower Sonoran near water.

Skunks forage slowly and clumsily in the dark for insects, berries and grasses, carrion, bird eggs, and rodents. They thoroughly chew bees and spit out the exoskeletons. With their strong paws and long claws, skunks scrape the ground for grubs and roots and can dig their own dens. Females raise four to eight kits by themselves, leading them on nocturnal expeditions when they are two months old. Great horned owls, which have a feeble sense of smell, are their primary predators. Striped skunks are common on the South Rim, while spotted skunks are the most common carnivores in the inner canyon's Lower Sonoran Zone.

DORLING KINDERSLEY

Squirrel, Abert's *Sciurus aberti.* Diurnal: ponderosa forest.

Tassel-eared squirrels bound across the forest floor and leap between trees balanced by plumed tails they use to shade themselves when resting. They court in spring, chasing each other up and down trees. Females bear hairless two-inch-long (5 cm) babies with closed eyes and ears, tiny claws, and fine whiskers in nests of twigs roofed with pine needles and lined with grass, leaves, fur, and feathers. Tassel-eared squirrels eat buds and male pinecones with nutritious yellow pollen and snip off twigs to eat the inner bark. In the rainy season they eat mushrooms and false truffles, a beneficial fungus on ponderosa roots that supplies water and nutrients in exchange for carbohydrates. The charcoal-gray Kaibab subspecies isolated on the North Rim has a coal-black belly and silvery tail. Common and widespread, Abert's squirrels are dark gray with reddish backs, white underparts, and white tails flecked with black.

NATIONAL PARK SERVICE

Squirrel, rock *Spermophilus variegatus.* **Diurnal: all habitats except grassland and riparian.**

C. Hart Merriam wrote that rock squirrels are the "most characteristic mammal of the piñon belt." Although large with long bushy tails like tree squirrels, they are ground squirrels closely related to antelope squirrels, chipmunks, and prairie dogs. Female rock squirrels burrow under rocks in the middle of large colonies; males live around the edges. They forage during the day for seeds, buds, leaves, and cactus fruit and brazenly beg from humans.

NATIONAL PARK SERVICE

Birds

BRIAN GATLIN

Bluebird, western *Sialia mexicana.* Summer: ponderosa forest, pinyon-juniper woodland. Winter: all other habitats.

Western bluebirds nest in tree cavities, especially those facing east at forest openings. In spring males are a deep indigo with rusty-colored vests. Flocks move across the landscape, perching on dried stalks or low branches, and dropping to forage for insects on the ground. When winter snows make this difficult, they move to lower shrubby places where they seek out seeds and berries.

BRIAN GATLIN

Chickadee, mountain *Poecile gambeli.* Year-round: coniferous forests, Upper Sonoran.

Mountain chickadees are one of the most charming and easy-to-identify birds on the rims of the canyon. Their distinctive call sounds like their name: chicka-chicka-dee-dee-dee. Tiny birds with conspicuous black caps, throats, and Zorro masks, they often join other small birds in mixed flocks foraging noisily from tree to tree for seeds and insects. They are curious and show little fear of humans.

BRIAN GATLIN

Condor, California *Gymnogyps californianus.* Designated an endangered and experimental/ nonessential population under the Endangered Species Act. Summer: nests in caves in the Redwall Limestone.

California condors are dark birds with nine-foot (3 m) wingspans, red heads without feathers that stay clean when dipped into dead animals, and bumpy tongues for holding slippery innards. Yet researchers know that despite their grim appearance and source of nourishment, condors are sociable and playful birds that often feed, bathe, and roost together. In courtship, pairs rock back and forth with their wings outstretched. Males neck-wrestle and push each other off perches. Condors have phenomenal eyesight but a poor sense of smell. Visual scavengers, they soar on the lookout for carcasses indicated by the presence of turkey vultures or ravens. During the Pleistocene, they fed on the remains of megafauna such as mastodons.

Creeper, brown *Certhia americana*. Summer: mixed coniferous forest. Other seasons: ponderosa forest, pinyon-juniper woodland.

Brown creepers have been described as "animated pieces of bark." They hitch upward on the trunks of conifers, spiraling around the trees probing for insects with their curved and sharply pointed bill. Their small size, mottled brown camouflage, and habit of hugging close to tree trunks make them difficult to see. They build nests of cocoons, moss, fibers, and feathers with an entry hole at the bottom and an exit at the top.

MIKE'S BIRDS/FLICKR

Dipper, American *Cinclus mexicanus*. Year-round: along creeks.

Dippers bob on rocks as they watch flowing water, then suddenly plunge in and swim or grapple along streambeds to catch insects and larvae. They build nests of moss and sticks behind waterfalls. A high level of hemoglobin in their blood, along with a slow metabolism and thick feathers, enables dippers to flourish in cold water. In winter they move down along the Colorado River.

NIGHTJAR09/DREAMSTIME

Dove, mourning *Zenaida macroura*. Summer: Upper and Lower Sonoran, coniferous forests.

Perched in open woodlands, mourning doves coo sweetly and a little sadly. They flutter down to explore the ground for seeds, which they store in their crop to digest later. Plump birds with small heads, they eat up to a fifth of their own weight each day. When disturbed they burst into flight with a clatter, their wings making a squeaky sound.

BRIAN GATLIN

BRIAN GATLIN

Falcon, peregrine *Falco peregrinus*. Year-round: all habitats (see page 55). Peregrine falcons mostly hunt birds and bats in flight at dusk and dawn. Tucking in their feet and wings, they dive down at speeds of up to 200 miles (300 km) an hour to stun their prey. Beautiful birds with a black head and moustache, blue-gray back, and barred chest, they are favored by falconers. They nest on cliff ledges near where they hunt; in turn, they are preyed upon by great horned owls and eagles.

THE GRAND CANYON FLYWAY

Now is the time of the great fall migrations, and in truth the whole world seems built for birds on the wing.

—Natalie Angier, *Songs and Sojourns of the Seasons*, 2007

Fall reveals another kind of "transition zone" along the canyon's rims. Each year thousands of raptors—birds of prey—pass over the Kaibab and Coconino Plateaus on their fall migrations from northern breeding grounds to the southern regions where they winter. The plateaus extend several mountain chains of the interior West farther south, lifting and guiding raptors to regions where warmer weather and longer days promise more insects, rodents, and other prey.

Along this route, at least sixteen species of hawks, eagles, and falcons take advantage of rising pockets of warm air called thermals, soaring upward and then gliding down into the next thermal, then soaring up again. Observers say that on a warm day, a hawk can travel 250 miles (400 km) without flapping its wings. On cooler days raptors use winds deflected upward by the plateaus and canyon walls, riding updrafts to travel great distances with a minimum of energy.

Most of North America's flyways have been known for a century, but the one over the Grand Canyon was only recognized in 1987. Observers from HawkWatch International monitoring raptor migration at Grand Canyon from 1991 to 2011 reported that the flight through this region is one of the largest concentrations of migrating raptors known in the western United States and Canada, with counts ranging from about 6,100 to 12,300 migrants per season.

HawkWatch observers also learned new information about the distribution and behavior of certain birds. For instance broad-winged hawks that breed in the western prairie provinces of Canada were once assumed to migrate eastward, then down to the Gulf Coast of Texas and along to Central and South America. However in 1998, at least thirty-five broad-winged hawks flew right over Lipan Point, likely on a direct route south from Alberta.

Finch, Cassin's *Carpodacus cassinii.* Summer: coniferous forest. Winter: Upper Sonoran.

BRIAN GATLIN

Cassin's finches nest in cups of sticks and lichens on branches, often less than 100 feet (30 m) from one another. During breeding season, males have a rosy cap that looks like a red toupee and mimic songs of other birds as well as singing their own. Cassin's eat mostly seeds, buds, and berries.

Flycatcher, southwestern willow *Empidonax traillii extimus.* Listed as endangered under the Endangered Species Act. Summer: riparian.

RICK FRIDELL

A neotropical migrant is a bird that breeds in North America and spends the rest of the year south of the Tropic of Cancer, an imaginary line in central Mexico. Having lost 90 percent of their riparian habitat, southwestern willow flycatchers are the most threatened of thirty or so neotropical migrants that spend part of the year at Grand Canyon. They perch on twigs and swoop out to snatch insects in flight or "hover-glean" them from leaves or the ground. Somewhat drab thicket-dwelling birds, they are most easily identified by their songs, which they sing with their heads thrown back in melodramatic display. They raise their little crests, fluff their chest feathers, and flick their tails at intruders, even other flycatchers.

Gnatcatcher, blue-gray *Polioptila caerulea.* Summer: pinyon-juniper woodland, Lower Sonoran riparian.

BRIAN GATLIN

Blue-gray gnatcatchers are tiny birds with long tails that they constantly flick from side to side, stirring up insects as they forage through foliage. During breeding season males have a thin black "eyebrow" and make a creaky sound.

Goldfinch, lesser *Spinus psaltria.* Year-round: riparian. Summer: mixed coniferous and ponderosa forests, pinyon-juniper woodland.

BRIAN GATLIN

Lesser goldfinches eat seeds of the sunflower family or buds of riparian trees, hanging upside down on the end of a stalk or twig that nods under their scant weight. They pair off to nest but are sociable the rest of the year. Two dozen may roost together in a single tree.

Females are olive colored; males are golden yellow with black caps, tails, and wings. Their high-pitched songs include a distinctive melancholy wheeze.

U.S FISH AND WILDLIFE SERVICE

Goshawk, northern *Accipiter gentilis.* Year-round: ponderosa forest.

Northern goshawks are an apex predator at the top of the Transition Zone food chain. Nicknamed "gray ghosts," they are reclusive except in spring, when males fly a flamboyant, undulating courtship display. Northern goshawks consume a great many small animals. Scanning the forest with keen red eyes, they dive suddenly from perches with a wing flap or fly low and fast through openings among the trees to ambush prey, especially tassel-eared squirrels.

BRIAN GATLIN

Grosbeak, evening *Coccothraustes vespertinus.* Year-round: ponderosa forest, pinyon-juniper woodland.

Evening grosbeaks are an irruptive species, common one year but often scarce the next. Although their long-term population at the Grand Canyon varies, in fall they usually migrate through in large numbers from elsewhere. Big, bright yellow birds with black and white wings, they have a sturdy bill for cracking nuts.

Hawk, red-tailed *Buteo jamaicensis.* Year-round: all open habitats.

Red-tailed hawks soar in circles, their russet tails fanned out for steering, their keen eyes scanning the ground for small mammals. Their numbers are linked to the amount of winter snow and rain; more green plants in spring mean more rabbits, increasing the hawks' success in rearing their voracious young. Red-tailed hawks glide down silently on their prey with outstretched talons.

BRIAN GATLIN

BRIAN GATLIN

Heron, great blue *Ardea herodias.* Year-round: aquatic.

River runners often see solitary great blue herons stalking the shallows of eddies and marshes, their daggerlike bills poised to stab at fish, toads, and frogs. At almost four feet (1.2 m) they are the tallest herons in North America, with long graceful plumes on their heads and chests.

BRIAN GATLIN

Jay, western scrub- *Aphelocoma californica.* Year-round: Upper Sonoran.

Like the other two jays at Grand Canyon, western scrub-jays are bold, noisy, sociable, and curious. They have gray backs, pale undersides, and bright blue caps, wings, and tails. They eat seeds—especially pinyon nuts and juniper berries—in fall and winter and rely more on insects the rest of the year. Courting pairs sing a long sweet duet, then stay together all year and join a flock in winter.

BRIAN GATLIN

Junco, gray-headed dark-eyed *Junco hyemalis.* Year-round: coniferous forest.

Dark-eyed juncos from different parts of the country vary in coloring and are considered separate "races" of the same species. Gray-headed juncos have a patch of rusty-red on their backs. They nest beneath bunchgrasses or tucked under rocks on slopes. In September, Oregon juncos with black hoods, cinnamon backs, and white bellies arrive at the Grand Canyon and stay until spring. Slate-colored juncos mostly migrate through but sometimes stay for the winter.

TATIANA GETTELMAN

Kestrel, American *Falco sparverius*. Year-round: all habitats.

American kestrels are the smallest and most colorful falcons in North America, with patterns of dark spots and bars on slate-blue, rusty-red, and white feathers. They nest in tree cavities in the forest, in crevices or on ledges in the inner canyon. They hunt mostly insects such as grasshoppers, as well as small snakes, birds, and mammals.

U.S. FISH AND WILDLIFE SERVICE

Kingfisher, belted *Ceryle alcyon*. Year-round except summer: aquatic.

Belted kingfishers perch or hover above open water watching for fish and anything else wriggling beneath the surface. Their long, heavy bills and huge heads with shaggy crests give them an alert and intelligent appearance. They leave the Grand Canyon to breed farther north.

BRIAN GATLIN

Nuthatch, white-breasted *Sitta carolinensis*. Year-round: ponderosa forest, pinyon-juniper woodland.

White-breasted nuthatches honk as they hitch face-first down trees in coniferous forests, probing the bark for insects. They spend most of the year in pairs but in winter join mixed flocks, which are more efficient at finding food and watching out for predators.

Oriole, Scott's *Icterus parisorum*. Summer: Upper Sonoran.

Scott's orioles are bright mustard-yellow birds that breed in the Lower Sonoran. They depend on yuccas, consuming nectar from the flowers, picking insects from the leaves, and weaving nests of yucca fibers that they hang from the plant.

BRIAN GATLIN

Owl, great horned *Bubo virginianus.* Year-round: all habitats.

Nights at the Grand Canyon echo with the low hoots of a higher-than-elsewhere number of great horned owls, the most feared avian predators in North America. Their keen eyes are fixed forward, but extra neck vertebrae enable them to swivel their heads in either direction. Soft wing feathers muffle their approach as they swoop down on rodents, rabbits, reptiles, and peregrine falcons, as well as other owls.

U.S. FISH AND WILDLIFE SERVICE

Owl, Mexican spotted *Strix occidentalis lucida.* Listed as threatened under the Endangered Species Act. Year-round: primarily cavities and ledges in the Redwall Limestone (see page 55).

Mexican spotted owls nest and roost in cool, shaded places that are relatively inaccessible to predators. Spotted owls have dark eyes and brown feathers sprinkled in a pattern of white spots. They hunt mostly rodents, although they sometimes prey on bats and other owls. Their territorial call is a distinctive four-note hoot; females produce a nasal whistle.

NATIONAL PARK SERVICE

Owl, northern pygmy *Glaucidium gnoma.* Year-round: ponderosa forest, pinyon-juniper woodland.

Northern pygmy owls' staccato toots create consternation among the birds on which they prey. Unlike other owls, they hunt in broad daylight, taking insects and small mammals as well as birds. Less than six inches (15 cm) long, they are grayish-brown with little white spots and bars, a barred white lower breast, yellow eyes and feet, and fully feathered legs.

NATIONAL PARK SERVICE

Raven, common *Corvus corax.* Year-round: all habitats except aquatic.

Two feet (0.6 m) long with a wingspan up to four feet (1.2 m) across, ravens appear solid black but their feathers refract ultraviolet light, which birds can see. Humans can only imagine how colorful ravens look to other ravens as they soar and wheel in sunlight. After a courtship of aerobatics, clucks, and

TATIANA GETTELMAN

THERMALS ALONG THE RIMS

No other birds in the world are so purely creatures of the air as are the swifts.

—Naturalist Kenn Kaufman, Arizona-Sonora Desert Museum

Warm air rising from the inner canyon creates thermal updrafts right next to the rims, an invisible habitat where we can easily see predators chasing prey as well as courtship and play. White-throated swifts flash by, making a loud staccato chatter and beating their wings in an alternating manner. Swifts spend most of their time in the air hunting insects, catching raindrops, and even mating in flight. Violet-green swallows also swoop and dart, snatching tiny juniper gnats and other insects buoyed aloft by thermals. Although the swifts and swallows belong to different orders of birds, they have developed similar characteristics in response to their environment—a process known as convergent evolution.

Ravens—endlessly inventive in their aerobatics—loop, fly upside down, and drop sticks to each other mid air. Kettles of turkey vultures spiral slowly aloft with their wings held up in a teetering V, often joined by a zone-tailed hawk or two.

California condors now soar in the thermals too. Most likely to be seen between March and July, they were reintroduced to the region in December of 1996 after being extirpated in the early twentieth century.

bill snaps, ravens mate for life. Grain, berries, maggots, baby birds, lizards, carrion…ravens eat them all. They often watch us, as they watch any animal for a chance to swipe what it's eating. Pairs collaborate; while one distracts a bald eagle, the other snatches its fish. Biologists call them kleptoparasites.

NATIONAL PARK SERVICE

Sandpiper, spotted *Actitis macularia.* Spring and summer migrant, possibly breeding: Colorado River.

Spotted sandpipers have orange bills and spots on their breasts only during the breeding season. It is the females that establish territories and court males in swooping and strutting displays, and the males that incubate the eggs and care for the young. Spotted sandpipers are very active birds, briskly tottering along riverbanks, lunging, probing, or snapping at prey from insects and aquatic larvae to amphipods.

BRIAN GATLIN

Solitaire, Townsend's *Myadestes townsendi.* Year-round: coniferous forest. Winter: pinyon-juniper woodland.

Townsend's solitaires are gray songbirds with long tails and a white ring around each eye. They nest in high-elevation coniferous forests, usually on the ground under a shrub. In autumn they move down to pinyon-juniper woodlands where they sing to establish territories and defend the juniper berries they subsist on all winter.

BRIAN GATLIN

Sparrow, white-crowned *Zonotrichia leucophrys.* Winter: Lower Sonoran.

White and black bands on their heads make white-crowned sparrows easy to identify. They spend the winter in the Grand Canyon's lower elevations hopping about foraging for insects in brushy areas. Typically, they migrate north to breed in spring but occasionally nest at high elevations in the Southwest.

BRIAN GATLIN

Tanager, western *Piranga ludoviciana.* Summer: coniferous forest.

Male western tanagers in breeding plumage are breathtaking. They have red heads, black backs, yellow bodies, and black-and-white wings. They nest and sing in the forests on both rims, then depart in fall to winter in the forests of Central America.

BRIAN GATLIN

Thrush, hermit *Catharus guttatus.* Summer: coniferous forest.

Hermit thrushes nest in shrubs or the lower branches of trees along forest openings. They rummage through leaf litter and shake grasses with their feet in search of insects and seeds. Like other birds they have a forked windpipe called a syrinx that can produce more than one note at a time, with which they sing lovely though wistful melodies.

BRIAN GATLIN

Titmouse, juniper *Baeolophus griseus.* Year-round: Upper Sonoran.

Juniper titmice are little gray birds with a tuft of feathers on their heads and a medley of songs and calls. They hang upside down to peck insects from tree bark and leaves. Females staunchly defend their nests of grass, shredded juniper bark, and hair, standing firm and hissing like gophersnakes at intruders.

Turkey, wild *Meleagris gallopavo.* Year-round: coniferous forest.

Male wild turkeys gobble and squawk from late March to early June, gathering harems of females in coniferous forests and high meadows. After breeding, the toms are solitary and hens nest in the forest. After the poults hatch, females and young form flocks and forage in meadows for

SCOTT ROOT/UTAH DWR

BRIAN GATLIN

insects. As acorns and pine nuts ripen in fall, wild turkeys move where ponderosa, Gambel oak, pinyon, and juniper trees mingle. Occasionally, turkeys are seen at lower elevations and have even been known to nest at Indian Garden.

Vulture, turkey *Cathartes aura.* Summer: all open habitats.

Turkey vultures lack the ability to sing and instead produce a very loud hiss that sounds like they're breathing through a scuba mask. They find carrion to eat by using their exceptionally keen sense of smell. In late summer they assemble in large "kettles"—teetering in circles with their wings held up in a shallow V—before migrating south.

BRIAN GATLIN

Warbler, Grace's *Setophaga graciae.* Summer: coniferous forest.

Grace's warblers nest and forage for insects at the tops of mature ponderosa and mixed coniferous trees. Although they are common and have bright yellow faces and throats, they are often heard but seldom seen because they stay in the treetops. Grace's warblers are so wedded to high elevations that as they migrate south to winter in the forests of Mexico, they avoid stopping in low deserts along the way.

BRIAN GATLIN

Woodpecker, hairy *Picoides villosus.* Year-round: coniferous forest.

Hairy woodpeckers lean back on their tails and hop up tree trunks with both feet at once, probing for insects. They are attracted to beetle-infested forests where their nestlings will thrive as they feast. Hairy and downy woodpeckers are both black and white and look much the same. But hairies are larger and louder and have longer, stouter bills. Downies have obvious tufts above their thornlike bills.

BRIAN GATLIN

Wren, canyon *Catherpes mexicanus.* Year-round: all inner canyon habitats.

The sweet cascading song of canyon wrens may be the most evocative sound of the Grand Canyon. Canyon wrens are well-suited to their stony habitat, with big feet and claws that grip bare rock. They appear to get all the moisture they need from the insects and spiders they eat.

ESCAPE TERRAIN

I was amazed that it showed little inclination to desert the rocky crag on which I surprised it. Both the [Mexican spotted] owl and I were startled; the owl because of the unprecedented interruption of the solitude of Horn Creek, and I because my footholds and handholds were very precarious.

—Geologist John Maxon, *Grand Canyon Nature Notes*, October 1932

Grand Canyon's exceptional steepness, depth, and ruggedness create a large area critical to certain wildlife. These animals live where they do not because of what they eat or the temperatures they can tolerate. They live where they are most safe from predators.

Mexican spotted owls nest in caves and on rocky shelves in steep side canyons where the great horned owls that prey on them have difficulty maneuvering. Surveys in 2001 and 2002 found over fifty Mexican spotted owl territories, indicating a thriving population of Mexican spotted owls within the park.

At least one hundred pairs of peregrine falcons nest on the Grand Canyon's cliffs where their eggs and nestlings are reasonably safe from foxes and coyotes but must be defended from ravens and owls. Rock wrens pave a little path with pebbles to their softly woven, cup-shaped nests in inaccessible crevices. They require no water, getting moisture from the insects they consume.

Canyon mice have large ears and a tuft of hair at the end of their tails. They inhabit bare rock where predators are few, dashing up sheer and sometimes overhanging stone in search of the few sparse plants or insects they might find to eat. Unlike most "mousey-brown" mice, they are yellowish and considered colorful. They drink little water, getting most of what they need from their food or entering torpor until the rainy season.

Rock pocket mice live among jumbled rocks to frustrate the efforts of owls to catch them. A genetic trait enables each population to develop coloration that matches the rocks among which they live.

Ringtails move about in the dark, slipping up and down cliffs and among rocks at all levels of the steep canyon walls to pounce on mice or ransack campsites. They favor places where the bobcats, coyotes, foxes, and owls that hunt them are at a disadvantage.

Male bighorn sheep stay above about 4,500 feet (1,400 m) from October through early March, when they migrate to lower elevations for the lambing season. Although we humans may consider it strange to go where it's so hot for the summer, male bighorns move down to the Colorado River or pools and waterfalls of the deep inner canyon to be near water. Their specially-adapted hooves can grip bare rock and cut into ice, enabling them to outmaneuver mountain lions.

Mule deer summer on the forested rims where it is cool and they can fatten on plenty of forage. There they frequent areas busy with humans, which predators avoid, or steep or rocky terrain where their stotting—stiff-legged bounding—gives them an advantage over less-nimble mountain lions, coyotes, and bobcats. When winter snows make it harder for them to flee, mule deer move down to rough country where they can dodge obstacles that slow down their pursuers.

Reptiles and Amphibians

KOJIHIRANO/THINKSTOCK

Chuckwalla *Sauromalus ater.* Diurnal: Lower Sonoran, canyon riparian.

Chuckwallas are up to sixteen inches (0.4 m) long with loose, wrinkly skin, thick tails, and wide, flat bodies. Males are black with speckled midsections, while females are brown with red spots. Chuckwallas are shy vegetarians that eat flowers and spend hours basking in the sun. When threatened, they wedge themselves into rock crevices by inflating their bodies.

NATIONAL PARK SERVICE

Frog, canyon tree *Hyla arenicolor.* Mostly nocturnal: inner canyon riparian.

In spring, male canyon tree frogs chorus to attract females in deep, rhythmic bleats that sound almost like sheep. Females lay eggs in water, where tadpoles eat algae and organic debris. Adults are mostly terrestrial, eating spiders, beetles, ants, centipedes, and other insects.

NATIONAL PARK SERVICE

Lizard, common collared *Crotaphytus collaris.* Diurnal: Upper and Lower Sonoran, canyon riparian.

Collared lizards are splendid reptiles that sometimes rear up and run on their hind feet like miniature tyrannosauruses. Males have turquoise and yellow bodies, black and tan rings around their necks, and long tails. Females are light brown. Aggressive predators of insects, small mammals, and other lizards, they will bite the hand that grabs them. Where common collared lizards come into contact with Great Basin collared lizards, they apparently hybridize.

DAVEFOC

Lizard, common side-blotched *Uta stansburiana.* Diurnal: Upper and Lower Sonoran.

Side-blotched lizards may have chevrons, stripes, blotches, or nothing on their brown, orange, or tan bodies. Consistently, however, they have a single dark blotch behind each foreleg. The three forms of males have different mating strategies: orange-throated are big and feisty, while blue-throated form stronger bonds with females.

Yellow-throated males resemble females, which they approach without provoking their orange-throated mates and engage when those mates are distracted.

NATIONAL PARK SERVICE

Lizard, desert spiny *Sceloporus magister.* Diurnal: Upper and Lower Sonoran.
Desert spiny lizards are named for their large, pointed, prickly scales. They are stout with big—sometimes orange—heads that have thin dark lines from their eyes to their necks. As with many lizards, their tails detach from their bodies easily when grasped by predators, but eventually grow back. Males have blue-green throat and belly patches.

NATIONAL PARK SERVICE

Lizard, greater short-horned *Phrynosoma hernandesi.* Diurnal: ponderosa forest, Upper Sonoran.

Greater short-horned lizards hatch inside their mothers. When tiny and smooth, they are coated by dust and match the dirt. Adults have "horns" and markings like litter on the forest floor. Slow and heavy-bodied, when approached they hold their ground with their fearsome spiky appearance and, if necessary, squirt a thin stream of blood from their eyes.

U.S. GEOLOGICAL SURVEY

Lizard, northern whiptail *Cnemidophorus tigris.* Diurnal: Upper and Lower Sonoran.

Young northern whiptails have stripes and thin blue tails that can be twice as long as their bodies. With age, their stripes break up into patterns of spots. Unlike other whiptails, males have black on their throats and chests. Northern whiptails live in open scrubby areas where they restlessly forage for insects, spiders, and scorpions.

PARAFLYER

Lizard, plateau *Sceloporus tristichus.* Diurnal: ponderosa forest, pinyon-juniper woodland.

Plateau lizards have coarse scales with prominent central ridges and are generally light brown with darker markings. Males have bright blue patches on their throats and a long blue smudge on each side that they flash by doing "push-ups" with their front legs.

U.S. FISH AND WILDLIFE SERVICE

Snake, gopher- *Pituophis catenifer.* Diurnal/nocturnal: all habitats.

Though sometimes active in daytime, gophersnakes are nocturnal when the weather is hot. Thick skin protects their noses when they shove their pointed heads into burrows after prey, which they find through their sense of smell. They constrict to suffocate rodents, lizards, and smaller snakes, and twine up trees to swallow bird eggs. When threatened they mimic rattlesnakes by flattening and raising their heads, hissing, and shaking their tails, but they are harmless to humans.

KCMATT1/BIGSTOCK

Snake, California king- *Lampropeltis getula californiae.* Primarily nocturnal: Upper and Lower Sonoran.

California kingsnakes are constrictors that prey on other snakes, small mammals, and birds. They are harmless to humans but may excrete an unpleasant musk when handled. Glossy black with narrow white bands that spread out on their bellies, they can be over four feet (1.2 m) long.

NATIONAL PARK SERVICE/FRANK WALLANDER

Snake, prairie rattle- *Crotalus viridis.* Primarily nocturnal and crepuscular: Upper and Lower Sonoran, often in rocky outcrops.

Prairie rattlesnakes are thick snakes up to five feet (1.5 m) long, with large triangular heads and blotches running down their backs. Their bodies vary in color, usually matching local rocks. They use heat-sensing pits between eye and nostril to hunt rodents, rabbits, birds, lizards, and other snakes, injecting them with a digestive venom and swallowing them whole. Grand Canyon rattlesnakes are a tan-to-pinkish subspecies (*abyssus*) found only at Grand Canyon.

NATIONAL PARK SERVICE

Toad, red-spotted *Bufo punctatus.* Nocturnal: riparian.

Red-spotted toads have pointed snouts, big round eyes, and grayish-tan skin covered with little red bumps called tubercles. They breed in creeks and pools. Females deposit gelatinous eggs that hatch in three days into tadpoles that transform into toads in six to eight weeks. In dry conditions the toads rest in cool, moist places such as burrows or under plant litter, emerging in large numbers when it rains.

Fish

Chub, humpback *Gila cypha.* Listed as endangered under the Endangered Species Act. Aquatic.

U.S. GEOLOGICAL SURVEY

The humpback chub is a streamlined fish with a concave head, long hump between head and dorsal fin, and deeply notched tail—adaptations that keep it upright and nimble as it pursues invertebrates in the Colorado River's strong currents. It spawns in warm, quiet backwaters, mostly near the confluence of the Little Colorado and Colorado Rivers, where the young remain until able to cope with the main river. Listed as endangered due to the effects of Glen Canyon Dam on the Colorado River, including colder temperatures, decreased turbidity, altered water chemistry, and the disruption of seasonal flows, as well as by competition and predation from non-native fish, they have begun to recover through translocations, modified operation of the dam, removal of trout around the Little Colorado River, and warmer water temperatures due to low lake levels in the first decade of this century.

U.S. FISH AND WILDLIFE SERVICE

Dace, speckled *Rhinichthys osculus.* Aquatic.

Speckled dace are the smallest native fish in the Grand Canyon, but also the most numerous. They have a black stripe from their

DESERT POTHOLES

Desert potholes are small, shallow basins eroded in bare rock. They fill with water after rainstorms and flash floods, then gradually dry up again.

Tiny crustaceans inhabit these pools, emerging from eggs for a brief life cycle of eating, swimming, and reproducing while there is still water in the pothole. They eat algae, bacteria, and microscopic organisms as well as dead things they find in their pools. Fairy shrimp are tiny, delicate crustaceans that swim on their backs, propelled by eleven pairs of legs. Clam shrimp are an ancient form of life with two-part shells that they close when threatened. They swim by sweeping with their antennae and use their legs to wave food into their mouths. Tadpole shrimp are another ancient life-form. That look like miniature horseshoe crabs with broad shells and narrow abdomens.

Potholes that are large enough may hold water striders, bugs with four legs long enough to distribute their body weight so they can walk on water. They may also support the tadpoles of Woodhouse's toads and canyon tree frogs. Although potholes are the most temporary of Grand Canyon habitats, their inhabitants go through all the stages of a highly evolved existence and are perfectly suited to their miniature world.

snouts through their eyes and an orange band on the sides of their speckled yellow bodies. They spawn in tributaries and are common where the river's flow is interrupted in eddies, riffles, side streams, and along the riverbank.

JOSEPH TOMELLERI

Sucker, flannelmouth *Catostomus latipinnis*. Aquatic.

One of four species of sucker once native to the Colorado River in the Grand Canyon, flannelmouth suckers have blubbery mouths with fleshy projections they use to suck up algae and aquatic insects. They spawn over gravel bars; when their young hatch they drift to backwater nurseries. They suffer the same threats as the humpback chub and other native fish and are candidates for threatened or endangered status.

U.S. FISH AND WILDLIFE SERVICE

Trout, rainbow *Oncorhynchus mykiss*. Aquatic.

Rainbow trout are native to cold, clear lakes and rivers west of the Rocky Mountains. Highly prized by sport fishermen for their "fight" and palatability, they are blue- or yellow-green with speckled backs and a streak of pink along their sides. Since completion of Glen Canyon Dam in 1963, the Colorado River below the dam has run cold and clear and is no longer subject to seasonal floods, thus creating prime habitat for rainbow trout. The Arizona Game and Fish Department stocked rainbows in the fifteen-mile (24 km) Lees Ferry reach between the dam and the boundary of Grand Canyon National Park; it has become a "blue ribbon fishery" for rainbows. In 1981 the Bureau of Reclamation implemented a flow regime for the dam that increased their size and numbers. However downstream of Lees Ferry, the trout compete with native fish for food and prey on young fish, including the endangered humpback chub, especially in their spawning area around the confluence of the Colorado and Little Colorado Rivers.

Invertebrates

SWEETCRISIS/BIGSTOCK

Bee, carpenter *Xylocopa* sp. Diurnal: Upper and Lower Sonoran.

Carpenter bees are very large, shiny black bees that visit the flowers of shrubs, transferring pollen stuck to their hairy back legs from one flower to another. Females bore a round tunnel several inches long into wood or agave stalks, push a ball of pollen to the far end, deposit an egg, seal it off with sawdust, then repeat until the tunnel is packed with egg chambers. Male carpenter bees behave agressively but have no stinger. Females do have a stinger that they will use if harassed.

Beetle, darkling or pinacate *Eleodes* sp. Crepuscular/ nocturnal in summer; diurnal in fall. All habitats.

U.S. FISH AND WILDLIFE SERVICE

Hikers often see a big black beetle bumbling across the trail and pause to investigate it. When the beetle senses them it halts, puts down its head, tips up its abdomen, and may eject a foul-smelling fluid that can irritate the pursuer's eyes or mouth. This habit has led to its nickname of "stinkbug." *Pinacate* is its name in the Nahuatl language of Mexico.

Cicada *Platypedia putnami.* Diurnal/crepuscular: pinyon-juniper woodlands, ponderosa and mixed coniferous forests.

VALERIY KIRSANOV/THINKSTOCK

In May and June, visitors may hear a faint clicking from conifers —the sound of cicadas of a genus nicknamed "wing-bangers" that tap their wings against needles and branches to attract mates. Emerging as larvae from the soil, cicadas pupate into adults with prominent eyes and transparent wings. They leave the husks of their larval stage clinging to trees. Females lay eggs in the bark of twigs, which hatch into nymphs that burrow into the ground where they feed on the juice of roots.

Dragonfly, flameskimmer *Libellula saturata.* Diurnal: aquatic/ riparian.

FRANCO FOLINI

Every part of a flameskimmer dragonfly is orange, including its legs, eyes, and wing veins. Up to three inches (8 cm) long, dragonflies are rapacious predators. They perch on grass or flower stalks and zoom out to grasp soft-bodied flying insects in basketlike jaws. Their larvae, called nymphs, live in the mud of streams or pools and prey on tadpoles or the larvae of other aquatic insects. Hikers may see flameskimmers anywhere near water, such as springs along the Tonto Trail.

Fly, spiny tachinid *Paradejeania rutilioides.* Diurnal: ponderosa forest, pinyon-juniper woodland.

WHITNEY CRANSHAW

There are over one hundred fifty thousand species of flies, from mosquitos to gnats to houseflies. Although some might consider all of them pests, many are not only beneficial but beautiful. Nicknamed "hedgehog flies," spiky tachinid flies have rounded amber abdomens with ruffs of stiff hairs. They are important late-summer pollinators of many flowers, especially those in the aster family.

WHITNEY CRANSHAW

Gnat, juniper *Culicoides* sp. Diurnal: pinyon-juniper woodland.

In early summer, strollers along the rims may experience an invisible irritant: tiny biting midges called juniper gnats. Larvae hatch from eggs laid in juniper bark and eat algae and microscopic bits of plants their first year. Clouds of adults emerge the next spring. Like their relatives the mosquitos, they pollinate flowers while consuming nectar, and only the females bite.

R. DODSON/BIGSTOCK

Scorpion, Arizona bark *Centruroides sculpturatus*. Nocturnal: desert riparian.

The bark scorpion is the most venomous scorpion in North America and the only scorpion at Grand Canyon that is dangerous to humans. Light brown and up to three inches (8 cm) long, it hides during the day under loose cottonwood bark and emerges at night to hunt big insects such as crickets. Care should be taken when handling gear and boots that have been left in shade or darkness.

NATIONAL PARK SERVICE

Tarantula *Aphonopelma behlei*. Primarily nocturnal: ponderosa forest, pinyon-juniper woodland.

Tarantulas are dark brownish-black, with two large round eyes with three smaller eyes on either side and two claws at the tip of each leg. They hunt insects at night by detecting their vibrations with hairy pads on the bottom of each foot. Male tarantulas prowl the forest in fall for females that attract them with pheromones. Although not dangerous, a tarantula's bite is painful, and bristles on their abdomens can irritate the skin.

NATIONAL PARK SERVICE

(Wasp) velvet ant *Dasymutila* sp. Diurnal: sandy soils.

Velvet ant females look like big hairy ants with fright wigs on their abdomens but they are actually flightless wasps. They scurry away from curious hikers although they may sting if handled. Males have wings and fly but are seldom noticed. There are about four hundred species of velvet ants in the American Southwest. Magnificent velvet ants are bright red, while thistledown velvet ants have bushy white hair.

INDEX

TOM BEAN

ABOUT THE AUTHOR

Susan Lamb served four years as the Desert View ranger-naturalist at Grand Canyon National Park and has since written two dozen books about the natural and human history of the American Southwest. She feels fortunate to live in a time when the observations of naturalists—once informed only by personal encounters with the natural world—have been borne out by the advanced techniques of modern science. Susan lives with her husband, photographer Tom Bean, in the ponderosa forest near Flagstaff, Arizona.

ACKNOWLEDGMENTS

I am grateful to the scientists and naturalists who cheerfully shared their research with me. Dr. Larry Stevens offered insights into the intricate realm of invertebrates, while Chuck LaRue brought his legendary powers of observation and recall to his review of the draft. John Coons contributed his deep knowledge of birds and tracked down researchers Steve Emslie and Larry Coats for details on woodrats. Don Lago's easy familiarity with Grand Canyon history informed my approach to John Muir, John Wesley Powell, and other historic figures.

I owe much to Grand Canyon National Park staff including RV Ward and Lori Rome, Brandon Holton, Janice Stroud-Settles, Greg Holm, Todd Chaudhry, Steve Rice, and Betty Upchurch. Kim Besom guided me through decades of wildlife notes, photographs, and specimens in the Museum Collection.

It has been a privilege to work with Pam Frazier, an exceptionally talented and considerate editor. I thank Amanda Summers for her creative and effective design.